Isah Haruna, Maigari, Isa, Mamman, Gusikit

Trace Elements Evaluation for some mine ponds in the Tin Mining Areas of Jos amd Environs, Plateau State, Nigeria

GRIN Verlag

Bibliografische Information der Deutschen Nationalbibliothek:

Die Deutsche Bibliothek verzeichnet diese Publikation in der Deutschen National-
bibliografie; detaillierte bibliografische Daten sind im Internet über http://dnb.d-
nb.de/ abrufbar.

Imprint:

Copyright © 2012 GRIN Verlag GmbH
Druck und Bindung: Books on Demand GmbH, Norderstedt Germany
ISBN: 978-3-656-33140-7

TRACE ELEMENTS EVALUATION FOR SOME MINE PONDS IN THE TIN MINING AREAS OF JOS AND ENVIRONMENTS, PLATEAU STATE.

1Haruna, A. I, 2Maigari, A. S., 3Isa, M. T., 4Mamman, Y. D., and 5Gusikit, R. B.

1,2 and 3 Geology Programme, Abubakar Tafawa Balewa University, Bauchi, PMB 0248, Bauchi State, Nigeria.
4 Geology Department, Federal University of Technology, Yola
5 Geology Department, University of Jos

ABSTRACT

The study area ("Derelict Land") in Plateau State fall within Y coordinates 1039185m to 1114995m and X coordinates 452385m to 514845m on the Nigeria's LANDSAT MSS 2001. The approximate area covered by the project is about $3178.1km^2 2$ from the satellite image measurements. Six mining areas were involved (Jos-Bukuru, Rayfield, Sabongidan kanar, Bisichi, Kuru III and Barikin Ladi) in this research. Analysis of land use changes on the confirmed that mining is rapidly claiming most of land. Preliminary application of Remote Sensing and Geographical Information System (GIS) to study satellite image of the mining areas is an attempt to evaluate and characterize the mining areas based on the spectral signatures of mine ponds, inactive abandoned mine dumps and structural pattern of the areas.

The analysis of variance of the average trace element concentration from the six mining localities showed that Iron (Fe) is the only trace element that show universal anomaly in all the mining localities compared to WHO (2002) Standards for drinking (0.30ppm). Iron (Fe) anomaly range from 0.39-2.54ppm. Besides Iron (Fe), each mining locality has its unique trace elements anomaly. The overall analysis of variance between the average trace elements concentration in waters from the six tin mining localities compared to the World Health Organization Standards for consumable water confirmed anomalies of Fe, Pb, Cr, Co, Cu, Ni, Cd in waters collected from the mining localities which portrays environmental degradation as well as a tendency for pollution. The distribution of trace elements anomalies appeared to be consistent with increase in population and industrialization in the order; Jos > Barikin Ladi >Rayfield. The solution calls for intensive utilization of the joint technologies of Remote Sensing and Geographic Information System (GIS) to make effective evaluation of the areas and come up with a cost effective and constructive reclamation and re-utilization scheme which appeals to the environment. The scheme should be flexible enough to accommodate and convert minimum damage to environment as there is no use of land that is completely neutral to the environment

INTRODUCTION

The study area ("derelict land") on Jos, Plateau tin fields (Fig. 1&2; Plate 1) fall within Y coordinates 1039185m to 1114995m and X coordinates 452385m to 514845m. The approximate area covered by the mine ponds and mine dumps is about $3178.1km^2$ from the satellite image measurements (area shown by black rectangular lines on Plate 1). The six mining areas are at Jos-Bukuru, Rayfield, Sabongidan kanar, Bisichi, Kuru III and Barikin Ladi (Fig. 2; Plate 1).

This study is concerned with assessing the extent of environmental degradation on Jos, Plateau using Remote Sensing and Geographical Information System (GIS) techniques to locate the mine ponds in the mining localities.

Environment is the sum total of all the external condition that may act on organisms of community to influence its development or existence. The external condition of interest are the physical, chemical and biological as they relate to mining, processing, weathering and most importantly the changes imposed by these conditions. Ever since the origin of the earth, it has been subjected to changes. One of such activities that affect the environment and impose some changes in both physical and chemical composition of the environment is mining. Mining according to Adiuki-Brown (1999) is an act, process or work of extracting mineral of economic importance from their natural environments and transporting them to the point of processing and use. A mine is therefore an excavation made into the earth for the purpose of extracting mineral of economic value. It is not a potential Dam (e.g. Mine Pond). Mining has taken place in Plateau State for almost 100 years (1902 – 1982) and this has rendered derelict a land of about $325km^2$, out of the total of $8,600km^2$ of the Jos Plateau area (Alexander, 1985). Oxeham (1966) defined "derelict land" as land that has been damaged by extractive on other industrial processes such that in its existing state, it is unsightly and incapable of reasonable beneficial use". The Land Resource Development Centre in 1976 estimated that some $316km^2$ of the total $8,600km^2$ of the Jos, plateau area has been damaged by mining.

The mining on the Plateau has imposed the following structure on the estimated '$325km^2$ derelict land':

   a. Inactive and Abandoned Mines (Mine Ponds/Dams)

   b. "Lotto" Wells.

The Promulgation of the Mine Lands Act of 1946 paved way for the reclamation of mined areas on the Jos Plateau. Active reclamation began in 1964 by the Mine Land Reclamation Unit (MLRU) and the Forestry Department of the Ministry of Agriculture and Nation

Resources (ATMN, 1969). Other bodies involved in the reclamation activity included the mining companies and the Joint Consultancy Committee (JCC) on Mine Land Reclamation. The reclamation involved essentially the bulldozing of mound and filling in the ditches and flooded Paddocks and thereafter, planting the affected areas with eucalyptus tree seedlings in order to stabilize the soil and produce a good tilt. This is the so called "Level and Fill" method. Only 12.37km^2 of land was reclamed before the reclamation came to more or less a stand still due to the fact the "Level and Fill": method later proved extremely expensive to shoulder by the mining companies. By implication while the reclamation stopped, the mining continued and this annulled the initial impact of the reclamation. Alexander (1990) summarized the problems of reclamation as legislative, adverse economy and political influences. Hence the need to adopt a new reclamation strategy involving minimal cost with guaranteed immediate economic benefits. A key to this approach involves an understanding of the soil of the mined lands, existing land uses and soil- crop relationships.

Commercial tin mining started in Jos, Plateau about the year 1902 and in 1909, and the Niger Company was the only operating company, producing over 5,000 tonnes of cassiterite (Calvert, 1977). Tin mining expanded slowly due to poor transport facilities at the beginning. The construction of the Bauchi light railway in 1914 and an increase in the demand and the price of tin in the World market and due to the high requirement of tin for military use between 1914 and 1919, the production of tin increased (De kun, 1965). The production of tin further expanded with the arrival of a branch railway line at Kafanchan in 1927. This trend continues until the economic depression of the 1930s which caused a drop in the price of tin. Production however declined again after the war as a result of stiff competition faced by mining companies in the World market.

As the tin mining continued to grow, so did the mining settlements which are closer to the mining sites. At least about 50 of them were established over four decades (1902 – 1942). Some of the smaller labour camps are Sabongida kanar, Bisichi, Doruwa Babuje, Gana Ropp, Naraguta and Rayfield. The larger centers include, Barikin Ladi, Bukuru and Jos. By 1943, when Tin mining peaked with the production of 15,843 tons of Cassiterite, there were up to 7500 miners' employees. The population of Jos had then grown to about 18,000 people. Between 1942 and 1945 production reached 17,000 tonnes during the Second World War due to manufactures of armaments.

By mid 1985 large scale mining has visually collapsed, and small scale mining continued. Small scale mining is a continuous processes on the Plateau with even some of the reclaimed mine lands in most of the former mining areas being reworked using Lotto mining techniques

(Eziashi, 1998). The abandoned Lotto mines are now posing even greater environmental hazards

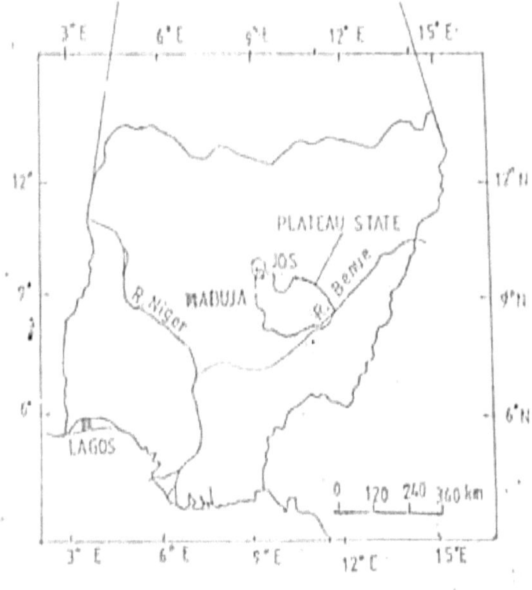

Fig. 1 *Location of the Study Area in Nigeria*

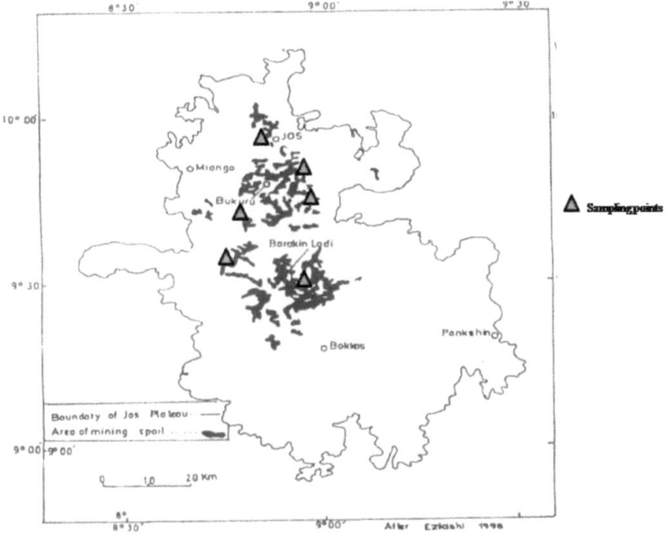

Fig. 2 Areas Affected by Tin Mining showing the sampling points on the Jos Plateau

(modified after Eziashi 1998)

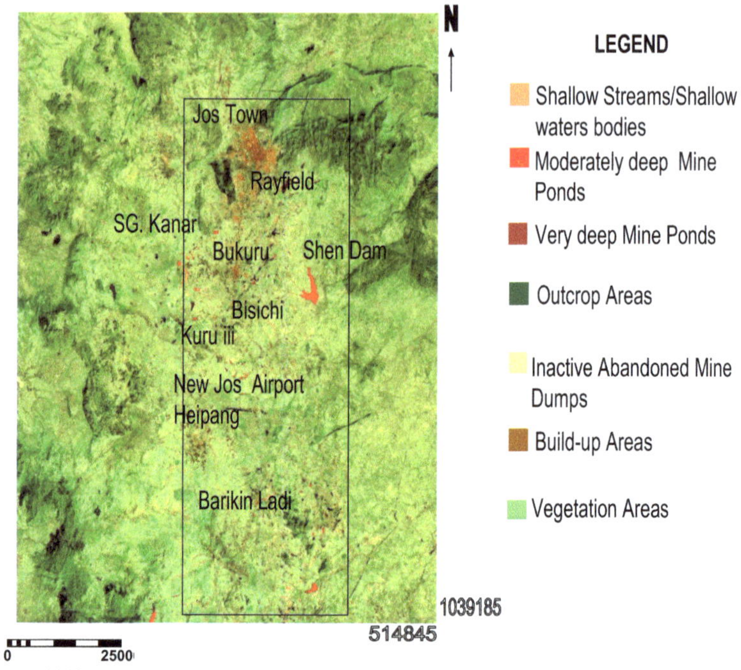

Plate 1. IMAGE OF MINING AREAS IN JOS AND ENVIRONMENTS FROM LANDSAT MULTISPECTRAL SCANNER (MSS) 2001

Geology of the Area

Jos Plateau is an isolated upland with an elevation of about 1219.20m above sea level and several satellite hill masses which rise from the surrounding plains (Macleod et al. 1971). The scenery varies from level plain and Plateau surface almost devoid of exposed rocks to rugged, deeply dissected massifs developed on the more resistant rock types (Macleod et al. 1971). The main rock types that exists on the Plateau are; the Basement Complex rocks including the Older-Granite and Metamorphic rocks (Gneisses and Migmatites), the Younger Granite from which Tin-ore (Cassiterite) originated and the Older and Newer Basalts (Fig. 3). The Biotite Granite of the Jurassic Younger Granite suites is the host for the Tin mineralization. The Tin mineralization is usually disseminated in the roof zone of the Biotite Granite Plutons and is related to post emplacement of hydrothermal greissenization, quartz veins and albitization (Macleod et al. 1971). The mineralized roof zone was weathered to form tin placer deposit along ancient river channels. Cretaceous basaltic flow (both Older and Newer basalt invaded the ancient river channels covering the tin placer deposits at their different periods of crystallization) and also the tin-bearing Biotite Granite were widely distributed in the surrounding drainage system (Macleod et al, 1970) (Fig. 3).

The research areas fall within the weathered basalts zone where the former and present legal and illegal mining took place (Fig. 2&3, Plate 1).

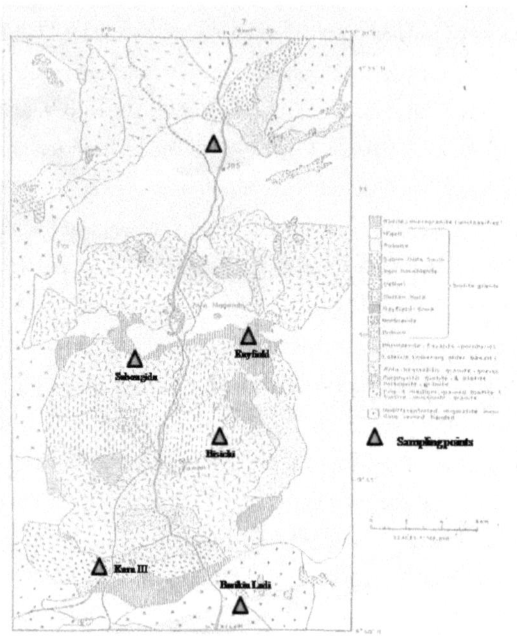

Fig. 3 Geological Map of the Mining Areas showing the sampling points.

(modified after Macleod 1971)

METHODOLOGY

The research entails the assessments of the impacts of mining on the environment through the generating chemical data from six mining localities within the areas rendered derelict by tin mining in Plateau State.

There are 2 sources of information as follows;

1. **Primary Source**
2. **Secondary Source**

The Primary Source involves acquisition of satellite image of the research area (LANDSAT ETM 2001) from the National Center for Remote Sensing in Jos. The Satellite image was used to identify the six mining localities using ILWIS software (which is Geographic Information System (GIS) software with image processing capabilities). All the six mining locations identified on the satellite image were visited on reconnaissance bases and three water samples were taken from the three mine ponds in each of the four mining localities. The water samples were analyzed for eight trace elements (Fe, Cu, Co, Pb, Zn, Ni, Cr and Cd) to generate part of the chemical data.

The Secondary Sources; the secondary sources involve collection of other related literatures and chemical data of the two other mining localities from other published works from the areas. The two chemical data would be merged, tabulated and averages of the trace elements concentration in the six mining localities would be computed using analysis of variance and then compared with the World Health Organization (WHO) permissible drinking standards to determine the tendency of pollution.

Results were used to propose remedy to problems detected by this research.

SAMPLE PREPARATION AND ANALYSIS

The twelve water samples from the mine ponds were collected (three from each locality) and acidified in sterilized plastic bottles to avoid core precipitation or reaction. The samples were subjected to partial digestion using dilute Nitric acid before the analysis. The mine pond water samples were analyzed using Atomic Absorption Spectrometer (AAS) and DR/2000 Spectrophotometer (Hach model).

RESULTS

The average results of the geochemical analysis from the four mining sites are contained on Tables; 1, 2, 3, and 5. The other two geochemical data generated from unpublished works are contained on Tables; 4 and 6. The overall averages results of geochemical data are contained on Table; 7 and it is the one that was used for the analysis of variance.

STATISTICAL ANALYSIS OF GEOCHEMICAL DATA;

A statistical test (Analysis of Variance) is carried out to test the variation in trace element concentration in water between the World Health Organization (WHO) permissible standard and the summation of average trace elements from the six different mining locations covered by Plate 1.

The two sets of data on which the decision were taken are: the average concentrations of trace elements (from six tin mining localities in Jos, Plateau State, Nigeria) and the World Health Organization (WHO, 2002) standard limit (trace elements in water) for drinking. The parameters on which decisions were taken were the similarity or variation of average trace elements from the mining areas with respect to the World Health Organization (WHO, 2002) standard limits which is the indices for determining pollution. A Statistical test was carried out to test the variation in trace elements concentrations between the World Health Organization (WHO) standards limits and the summation of average trace elements from the six different mining locations covered by Plate 1.

Statistical Test for Waters;

A paired test (null hypothesis and alternative hypothesis) is used to test whether or not there is a variation between the World Health Organization standard limit and the average trace elements concentrations in six mining localities. Variation in this context could either be positive or negative.

Positive Variation "means *an abnormality or a tendency for pollution (i.e. the magnitude of the average trace elements in mining areas is greater than the World Health Organization standard limits)".*

Negative Variation "means *normalcy or no tendency for pollution (i.e. the magnitude of the average trace elements in mining areas is less than or almost equal to the World Health Organization standard limits)"*

Null Hypothesis (Ho) = there is negative variation in magnitude between the average trace elements in the mining areas and the World Health Organization (WHO) standard limits of trace elements in drinking water.

Therefore **Ho**; WHO standard limits $\geq$ Average trace elements

(If this hypothesis holds, then that means there is no tendency for pollution)

Alternative Hypothesis (H1) = there is a positive variation in magnitude between the average trace elements in mining areas and the World Health Organization (WHO) standard limits of trace elements in drinking water.

Therefore **H1;** WHO standard limits $\leq$ Average trace elements

(If this hypothesis holds, then that means there is tendency for pollution)

Note that the equality sign is present in both hypotheses because it is the standard procedure to include the equality sign in Null Hypothesis.

The average concentrations of eight trace elements were calculated from three pond water s from asch of the four mining localities (Bisichi, Kuru iii, Rayfield, and Sabongidan Kanar) in this research and the average trace element data of two pond water samples (from Barikin Ladi and Jos mining localities) form other existing literatures were use as presented in tables 1-7.;

Table 1. Trace Element Mean of Three Pond Waters from Bisichi

Element Analysis in ppm	Bisichi			
	Minimum of three samples	Mean of three samples X	Maximum of three samples	WHO Standard 2002
Fe	0.14	0.39	0.81	0.30
Pb	0.01	0.32	0.04	0.35
Zn	0.17	0.23	0.31	5.00
Cr	0.01	0.017	0.02	0.05
Co	0.01	0.023	0.03	0.01
Cu	0.01	0.01	0.01	1.0
Ni	0.01	0.035	0.06	0.10
Cd	0.01	0.0015	0.02	0.01

Data from present work

Table 2. Trace Element Mean of Three Mine Pond Waters from Kuru iii

Element Analysis in ppm	Kuru III			
	Minimum of three samples	Mean of three samples X	Maximum of three samples	WHO Standard 2002
Fe	0.004	7.92	12.87	0.30
Pb	0.0388	0.066	0.1266	0.05 or -1.0
Zn	----	----	----	----
Cr	----	----	----	----
Co	----	----	----	----
Cu	----	----	----	----
Ni	0.012	0.033	0.0552	0.1
Cd	----	----	----	----

Data from the present work

Table3. Trace Elements Mean of Three Mine Pond Waters from Rayfield

Element Analysis in ppm	Rayfield			WHO Standard 2002
	Minimum of three samples	Mean of three samples X	Maximum of three samples	
Fe	0.01	2.54	9.7	0.3
Pb	0.01	0.041	0.2	0.05
Zn	0.01	0.11	0.32	1.05
Cr	0.01	0.018	0.03	0.05
Co	0.01	0.013	0.03	0.01
Cu	0.01	0.014	0.02	3.0
Ni	0.01	0.032	0.06	0.02
Cd	----	----	----	------

Data from present work

Table 4. Trace Elements Mean of Three Mine Pond Waters from Barikin Ladi

	Barikin Ladi			WHO Standard 2002
Element Analysis in ppm	Minimum of three samples	Mean of three samples X	Maximum of three samples	
Fe	0.10	1.69	3.83	0.3
Pb	0.004	0.031	0.09	0.05
Zn	0.006	0.041	0.10	5.0
Cr	0.06	0.11	0.20	0.05
Co	1.32	4.01	9.60	0.01
Cu	---	---	---	---
Ni	0.06	0.745	1.67	0.10
Cd	----	---	---	---

Unpublished B. TECH. Project, ATBU, Bauchi

Table 5. Trace Elements Mean of Three Mine Pond Waters from Sabongidan Kanar

Element	Sabon gidan Kanar			
Analysis in ppm	Minimum of three samples	Mean of three samples X	Maximum of three samples	WHO Standard 2002
Fe	0.3371	1.68	3.5926	0.3
Pb	0.0019	0.0020	0.0022	0.05
Zn	0.0180	0.035	0.0532	5.0
Cr	---	---	---	---
Co	0.022	0.0263	0.0338	0.01
Cu	---	---	---	---
Ni	---	---	---	---
Cd	---	---	---	---

Data from the present work

Table 6. Trace Elements Mean of Three Mine Pond Waters from Jos Mining District

Element	Jos mining District			
Analysis in ppm	Minimum of three samples	Mean of three samples X	Maximum of three sam[ples	WHO Standard 2002
Fe	0.04	0.83	1.7	0.3
Pb	0.00	5.0	1.5	0.05
Zn	19.7	19.25	59.0	5.0
Cr	0.00	0.67	2.10	0.05
Co	---	---	---	---
Cu	1.53	6.82	14.38	0.01
Ni	---	---	---	---
Mn	0.0021	0.019	0.061	0.01

Ogezi et al., 1985

Table 7. Average Mean Values (X) for Trace Elements (ppm) of Eighteen Mine Ponds Water Samples from Six Mining Locations, World Health Standards, and their differences

Elms.	Bisichi	Kuru III	Barikin Ladi	Rayfield	Sabon Gidan Kanar	Jos	Total Mean X	WHO Stds. 2002	D	D^2
Fe	0.39	7.92	1.69	2.54	1.68	0.83	2.51	0.30	2.21	4.88
Pb	0.02	0.07	0.03	0.04	0.002	5.0	0.86	0.05	0.81	0.66
Zn	0.23	----	0.04	0.11	0.04	19.25	3.28	5.00	-1.72	2.96
Cr	0.02	----	0.11	0.02	----	------	0.03	0.05	-0.02	0.0004
Cn	0.023	----	4.01	0.01	0.03	----	0.68	0.01	0.67	0.45
Cu	0.01	----	------	0.014	----	0.32	1.14	1.0	0.14	0.02
Ni	0.04	0.03	0.75	0.32	----	-----	0.19	0.01	0.09	0.008
Cd	0.002	----	----	----	-----	0.67	0.11	0.01	0.1	0.01
Mn	---	---	----	-----	-----	0.02	0.003	0.05	-0.05	0.0025
									∑D=2.23	∑D²=8.99

D	D^2
2.21	4.88
0.81	0.66
-1.72	2.96
-0.02	0.0004
0.67	0.45
0.14	0.02
0.09	0.008
0.1	0.01
-0.05	0.0025
∑D=2.23	**∑D²=8.99**

Using the formula

$$t = D/Sp$$

And $\quad Sp = \sum D^2 - (\sum D)^{2/n} / n-1$

Where t = Critical point

X = Mean values for the Trace Elements in water from mining localities

WHO= World Health Organization drinking standards

D = difference between the two set of data

D^2= square of the difference between two set of data

n = number of observations

So **n = 2 (W.H.O. standards and average values from mining locations)**

$\sum D = 2.23$

$\sum D^2 = 8.99$

Therefore

$SP = \sum D^2 - (\sum D)^{2/n} / n-1$

$= (8.99)^2 - (2.23)^{2/2} / 1$

$= 80.88 - 2.23$

$= 78.8901$

SP = 78.59

Calculate (t)

Using (t) tables

Degree of freedom = n – k

Where

n = number of observation

k = number of variables

Therefore

n = 2

k = 1

Degree of freedom = 2 – 1 = 1

Using 5% significant level

5% = 5/100 divide by 1

= 0.05

Therefore tabulated t = 0.05

Plot   t = D/SP

t = 2.23/78.59

Calculate t = 0.028

 Therefore

Calculate t = (0.028) < tabulated t = (0.05)

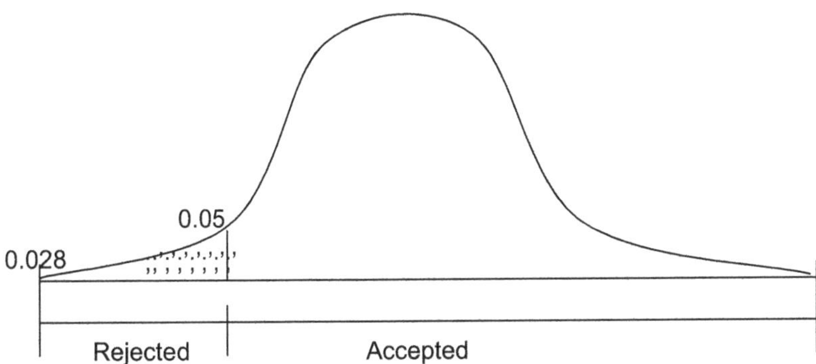

Fig. 5 Graphical expression of Calculated t and Tabulated t which favors Alternative Hypothesis

Decision

Since the Calculated t (0.028) is less than the tabulated t (0.05), the null hypothesis is rejected and the alternative hypothesis is accepted. In other words there is a Positive variation or a significant difference in magnitude between the World Health Organization standard limits of consumption of trace elements in drinking water and the anomalous average concentration of trace elements in the mining areas as shown by the graph (Fig. 5). Most of the averages trace elements in the mining areas are higher than the World Health Organization standard limits in drinking water as such there is a confirmed tendency for pollution and contamination of water in the mining localities in the Tin mining areas in Plateau State, Nigeria.

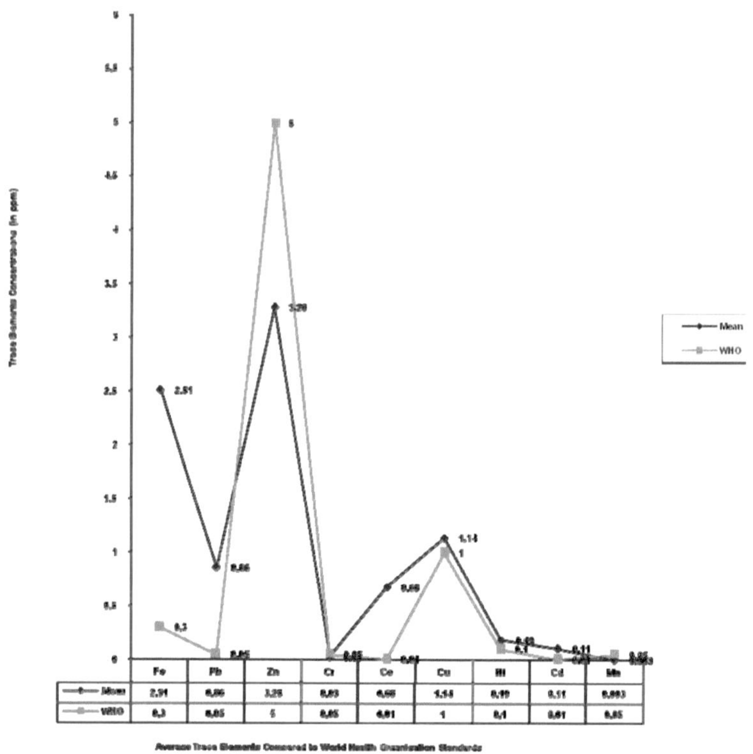

	Fe	Pb	Zn	Cr	Co	Cu	Ni	Cd	Mn
Mean	2.51	0.86	3.26	0.03	0.66	1.14	0.19	0.11	0.003
WHO	0.3	0.05	5	0.05	0.01	1	0.1	0.01	0.05

Fig. 6 Comparison of Average Concentrations of Trace Elements in the Mining Areas
With the World Health Organization Standards for Consumption in Drinking
Water

DISCUSSION

The results of the Analyses of Variance (ANOVA) conducted to test for the variation in concentration of Mean or Average Trace Elements in the Mine Ponds from six mining locations (Jos-Bukuru, Rayfield, Sabongidan kanar, Bisichi, Kuru III and Barikin Ladi) and the World Health Organization (WHO, 2002) permissible standards limit. The WHO, 2002 standard is also adopted as the baseline data because it is the threshold above which pollution is confirmed. In the averaging of result of water analyses from the six mining locations (Fig.2 Table 8) Fe 2.5ppm, Pb 0.86ppm, Co 0.68ppm and Ni 0.19ppm, all exceed the WHO permissible limit of Fe 0.03, Pb 0.85ppm, Co 0.01ppm and Ni 0.10ppm. All the six mining locations are consistent with respect to abnormal concentration of Fe (Table 7, Fig.2). Any value above the threshold marked by WHO 2002 standards is considered an anomaly in this research without any ambiguity to it's closeness to the WHO standard. This implies that trace elements are added to water bodies from mining activities and after the mining, finer fractions of the mine dumps also add trace elements to the mine ponds and dams.

Further detailed research could throw more light on this. The finer fractions of the fresh mine dumps or soil may contain the highest quantities of trace elements especially the clayey fractions. The mine dumps whether buried or on the surface will continue to undergo leaching to release the trace elements into adjacent mine ponds, streams and possibly the wells. This is how the mine pond waters and the streams become polluted. These same waters have been used as sources of water in some mining camps for domestic uses, irrigation, husbandry and industry. These automatically become the pathways through which the abnormal quantities of trace elements enter the human bodies through the food chain. Ogezi et al, (1999) reported that it is now widely recognized that the important factor determining the effect of a metal on organisms is not generally its total concentration but rather its speciation (i.e. the specific forms and nature of the elements). Metals may occur as hydrated ions, as complex inorganic or organic ligands, as absorbed and adsorbed ions, as colloids, as precipitation and in live and dead Biota.

Most trace elements either in water or food are very essential to human life. They are present in certain trace proportions in all healthy living tissues of all species and the deficiency symptoms may cause the malformation or malfunction of the tissues which would inhibit the normal development and growth of living tissues. Hence, in the studies of trace elements from the possible sources to man, the general trend followed is that an undersupply leads to element deficiency, sufficient supply result in optimum condition while oversupply result in toxic effects and possible dead in the end. When considered from environmental pollution point of view, metals may be classifies as follows:

a. **Non-critical**; such as Na, K, Mg, and Ca, which are essential metals and hard acceptors of electrons (Lewis acids) and which form stable bonds with harder donors (Lewis bases) such as H_2O, OH^-, Cl^- etc.

b. **Toxic**; but very insoluble or very rare, Ba, Ru, Ti, Hf, Zr etc which form weak bonds.

c. **Very Toxic** and relatively inaccessible metals e.g. Be Co, Sn, Ni, Ag, Pt, Pb, Zn, etc.

. Common very toxic members (As, Be, Cd, Cr, Cu, Pb, Zn, Hg, Mo, Se, Ag, Th, and Zr) in the mining areas World over become major health hazards. Some examples are the mercury poisoning in Minamata Bay, Japan. Cadmium poisoning causing the "itai itai" disease in the Japanse Villages on the banks of the Jintsu River Tomaya, Japan. Lead poisoning as in Leiping, Germany, during the summer of 1930 etc. (Forster and Wittman, 1983). A large number of such studies have been documented from Japan, Europe and U.S.A., but similar studies are yet to be published on Nigeria.

Because of the problem of pollution leading to poisoning (caused by mining, industrialization and utilization of chemical in stabilizing soils) a number of developed counties, have come up with variable national standards that determine the quality of water for different uses beside the one recommended by Who Health Organization. Although no such enforceable standards or environmental protection laws currently exist and practicable in Nigeria, the quality of Nigeria's surface and subsurface waters can be compared with these International Standards so as to establish concentration levels below which or above which harmful effects may result. In general, there are five different Major source from which metal pollution of the environment could originate, these are; Geological weathering of rocks, Industrial processing of Ores and Metals, the widespread use of metals and metals components, leaching of metal from domestic refuse, mechanical/smelting workshops and inactive abandoned mine dumps, and other human activities. Plateau State has several mineral occurrences, mostly associated to Igneous/metamorphic rocks and their weathering products, the State thus has a long history of mining, mineral processing and smelting. The general problem which arises in solving environmental problem is how to distinguish between metals added to the environment from natural geological weathering, mining and industrial activities and how or where they come from.

After the promulgation of the Mine Land Acts of 1946, the Forestry Department of the Federal Ministry of Agriculture and the Mining Companies commenced reclamation in

1949 by "Land and Fill" method and planting eucalyptus tree seedlings in order to stabilize the soil and produce a good tilt. About sixty (60) mining reclamation areas covering 1,237hectares of mine areas were established. However this figure (1,237hectares) was just a fraction of the total areas affected by mining and the reclamation came to a standstill (Dabi, 1999). The disadvantages of the "Land and Fill" method are; high cost implication, no adequate provision to take care of buried acidic mine dump soils and the acid mine waters that could be leached from them at depth and from fresh mine dumps on the surface. In essence Land and Fill method can only leveled the mine dumps and possibly cover the mine ponds but it could not take care of pollution as it was not meant to. A genuine reclamation plan supposes to address pollution in the environment besides the filling of mine ponds, ditches and pits.

CONCLUSIONS

The area rendered derelict by mining in Plateau State is about 315km^2. Patterson (1983) has reported that this area hosts 1000 mining ponds. "Derelict Land" is the inactive abandoned mine dumps, mine ponds/dams, and lotto wells. The satellite image proved that the areas affected by mining (Derelict Land) in Plateau State is far beyond the estimated 315km^2 declared by Patterson (1985).

The result of statistical analyses of different Mine Pond waters revealed contributions from different mining locations (Table 7) but the general contribution of trace elements appeared to be consistent with population and industrial growth. These industries could be mining or others but in any case they could be mechanical. The highest pollution was noted in Jos followed by Barikin Ladi and Rayfield. In rainy reasons, leaching of the granitic rocks is confirmed to add more trace elements to soil and water bodies. The rain water can percolate through joint and fractures in igneous and metamorphic rocks thereby leaching the trace elements and add them to the environment. The leached trace elements in the mine ponds/dams, streams, or "lotto" wells are concentrated in clays as hydrated ions, as complex inorganic ligands, as adsorbed or absorbed ions on colloids which may then be accessible to Man, Plants and Animals by either drinking such polluted waters and eating the plants and animals that drink or eat from these same polluted waters. Waters and food are the pathways through which the trace elements enter the food chain. During dry seasons because of excessive evaporation the mine ponds becomes highly concentrated in trace elements and these could extend to underground water by the process of osmosis.

A major negative effect of mining is the tremendous effect it has on land. This land is often a reservoir for water and could be use for Agriculture along present and ancient streams channels as in the case with the alluvial tin mining areas of the Jos Plateau. Mining also affect stream causes, divert and dam streams and pollute waters. Erosion and the deforestation process are hastened by the removal of soil and vegetation for mining. The water in boreholes, wells and mine ponds are actually interconnected through sub-surface permeable soils and fractured rocks. When the dilution factor is exceeded pollution is noticed in wells and boreholes. This level of pollution has started manifesting in some wells within the mining localities as shown by the chemical data (Table 7),

Detailed and accurate quantitative data on pathways of pollutants is lacking but a consideration of such pathways highlights processes that need monitoring For effective environment monitoring, an interdisciplinary approach and international cooperation are necessary since the process cut across disciplines and national boundaries and also affect the atmosphere and hydrosphere. Systematic use of Remote Sensing and Geographic Information System could simplify the task.

REFERENCES

Adepetu, A.A. (1985); Farmers and their Farms on four Fadamas on Jos Plateau. Jos Plateau Environment Resources Development Programme (JPERDP). Interim Report 2.

Adiuku-Brown, M.E. (1998) Effect if mining and Associated by-Products: A Special Study of Trace Elements of the Jos Plateau and Zurak Mine Dumps. North Nigeria. PhD Thesis in Preparation. Department of Geology and Mining, University of Jos, Nigeria.

Adiuku-Brown, M.E. and Ogezi, A.E (1984) Comparative Sample Decomposition Techniques for the Analysis of Sulphides (Full Paper – 10p + 12 tables).

Adiuku-Brown, M.E. and Ogezi, A.E. (1991); Heavy Metal Pollution from Mining Practices. A Case Study of Zurak. Journal of Mining and Geology Vol. 27(2) pp.205 – 211.

Adiuku-Brown, M.E. (1999); The Danger posed by the Abandoned Mine Ponds and Lotto mines on the Jos Plateau. Journal of Environmental Sciences. University of Jos. 3(2) pp.258-265.

Ajeabu, H.O. (1981); Mining and Land Degradation on the Jos Plateau. Paper Presented At the 24th Annual Conference of Nigerian Geographical Association. 20pp.

Akande, O.F. R., (2002); The Use of GIS And Remote Sensing Techniques in the Assessment of Land Use and Land Cover Changes in Kaduna South and Environs Between 1980s to 1990s. B.Sc Dissertation. Geography Department. Ahmadu Bello University, Zaria.

Aku, I.M. and Schoeneich, K (1998); <u>The Study of Degraded Mine Lands of Jos, Bukuru, Riyomi, Barikin-Ladi and Bokkos Areas of Jos Plateau for Development Possibilities,</u> Vol. 3. publish by Government of Plateau State. Environmental Protection Agency.

Alexander. M.J. (1984); Soil Development on Mine Spoil and Reclaimed Areas of the Jos Plateau: Soil Profile Description. Jos Plateau Environmental Resources Development Programme (JPERDP) Interim Report 1. pp78.

Alexander. M.J. (1985); A Historical Introduction to Reclamation of Mine Land on the Jos Plateau. Jos Plateau Environmental Resources Development Programme (JPERDP) Interim Report 4. pp 24.

Alexander. M.J. (1986); Soil Characteristic and the Factor Influencing their Development on Mine Spoil of the Jos Plateau. Jos Plateau Environment Resources Development Programme (JPERDP).

ATMN, (1969); The Plateau Mine Field. A Booklet Prepared by the Amalgamated Tin Mining Company of Nigeria. (ATMN).

Calvert, A.F. (1977); Nigeria and its Tin Fields. Edward Stanford, London

Dabi, D.D. (1990); Farming Possibilities on the Soil of the Heavily Mined Terrain in Rayfield Areas of the Jos Plateau. Unpublished M.Sc Thesis, Dept. Geography and Panning, University of Jos, Nigeria.

Dabi, D.D. and Nyagba, J.L. (1999); an Alternative Approach to the Reclamation of Mine Lands on the Jos Plateau, Nigeria. Journal of Environmental Sciences 3(2), 1999, pp 245-252.

De Kun, N. (1965); the Mineral Resource Africa. Elsevier. Amsterdam 483p.

Durkin, T. V. and Herrmann, J. G. (1994) ; Focusing on the Problem of Mining Wastes: An Introduction to Acid Mine Drainage. Reprint from EPA Seminar Publication no. EPA/625/R-95/007. "Managing Environmental Problems at Inactive and Abandoned Metal Mine Sites" Anaconda, MT, Denver. pp.4.

Eziashi, A.C. (1998); Barriers to the Suggestions towards Effective Management of the Degraded mine lands of the Jos Plateau. Journal of Environmental Sciences, (2), pp 41- 48

Fisher, W.A. 1995: History of Remote Sensing in Reeves R.G. (Ed) Manual of Remote Sensing American Society of Remote Sensing Falls Church.

Forstner, U. and Wittmann, G.T.W. (1983); Metal Pollution in the Aquatic Environment.2nd Ed. Springer. P.486

Gajera, E.N., Eguaroje, O.E. and Omomole (2004); <u>Fundamental of Remote Sensing and Image Interpretation.</u> A Workshop Report National Center for Remote Sensing, Jos.

Gyang, J.D. (2002); Assessment of Impart of Mining Activities on parts of Barikin–Ladi Areas of Plateau Sate. Unpublished Master Thesis submitted to Department of Geography, Federal University of Technology, Minna.

Kyari, J.D. (2001); Nature of Land –Use Changes and Land Degradation in Jos and Environs, Post Graduate Seminar Paper, A.B.U. Zaria.

Kyari, J.D. (2002); Application of GIS and Remote Sensing Notes. Unpublished, A.B.U. Zaria.

Lillesand, T.M. and Kiefer, R.W. (1979); Remote Sensing and Image Interpretation. John
Wiley and sons, New York.

Macleod, W.N., Turner, D.C. and Wright, E.P. (1971); The Geology of the Jos Plateau.
Vol. 1 General geology, Geological Survey of Nigeria Bulletin No.32. pp11-34

Morgan,W.T.W. (ed) (1979); The Jos Plateau: A survey of Environment and Land use.
Occasional Publication (New Series) 14, Dept. of Geography University of Durham,
U.K.

Ogezi, A.E., Adiukwu-Brown, M.E. and Alam, M. (1985); Environmental and Economic
studies
of Trace Elements associated with mining and mineral processing in parts of Plateau
State, Nigeria- A Preliminary Report. Pp 67-96

Ogunseyin, O. (2004); Application of GIS and Remote Sensing to Changes in Land use
in the Federal Capital Territory. B.Sc Dissertation, Geography Department,
Ahmadu Bello University Zaia.

Oxeham, J.R. (1966); Reclaiming Derelict Land. Faber and Faber London, U.K.

Patterson, G. (1986) ; Tin Mine Ponds of the Jos Plateau, their Nature and Resource
Value. JPERDP Interim Report No. 8. pp66.

Rose, W.A., Hawkes, H.E. and Webb, J.S. (1979) ; Geochemistry in Mineral Exploration.
2nd Edition, Academic press, New York. Pp 548-581

Stan, Aronoff, (1999); Remote Sensing and GIS Technology for Field Implication in the
Amazon, Science 260.

Gerstman, Bud (2003) Introduction to Hypothesis Testing (6), StatPrimer (c). An Online
Textbook on Statistics, Chapter 6, pp8.
(http/www.sjsu.edu/faculty/gerstman/StatPrimer/)

Strahler, A.N. and Strahler A.H. (1992); Environmental Geosciences. John Wiley and
Sons, Inc. 511p.

Ugodulunwa, F.X.O. and Taiwo, A.O. (1997); Reduction of Environmental Impacts of
Mining and Mineral Processing through Environmental Education. Environmental
Education for Sustainable Development: Focus on Nigeria by S.U. Udoh and G.O.
Akpa (eds) Education Books pp133-139.

Wolff, F. E., Mckay, Jr. D. T. and Norman, D. K. (2005); Inactive and
Abandoned Mine Lands. Washington Division of Geology and Earth Resources
Information, Circular 100. pp. 18.

World Health Organisation (WHO) (2002); Guidelines for Drinking Water Quality.